The Mysteries, Fun and Gratification of Ancestral DNA Testing

Who Are You?

I0467638

By: James M. Lowrance © 2015

TABLE OF CONTENTS

1. How My First Testing through Ancestry.com Compared to My Known Family History
2. My Reaction to the Analytical Ethnicity Maps
3. My Second DNA Testing through 23andme.com
4. My Complaint about "Southern Europe" and a 23andme Surname Affiliate Site
5. Lowrance History – Some Recent and Some Ancient
6. The Expected Changes coming soon to 23andme.com
7. Discussing Health Issues on the 23andme Forum
8. Do Names on your DNA Relatives Discovery List Change on Occasion (Question directed at me)
9. When I started with GedMatch.com
10. My Expressed Disagreement with 23andme.com over a Key Area Regarding Countries
11. My Favorable Review for 23andme.com

Who are You?

INTRODUCTION

This book (8,865 words in length) is not deeply scientific but rather relates the journey (mine) that a person will take generally who is seeking their ancestral history, including learning enough to understand how to interpret their DNA analysis. As I relate in my numbered posts/articles that follow, there can be satisfaction, surprises and at times, disappointments. All of this can happen when one takes a look at the results of ancestral DNA testing. I wrote this informational book in parts over several months, like one might make entries into a diary. What the information will do, is to help other people considering ancestry testing and analyses, to roughly know what to expect from them. Knowing your ancestral background can be tremendously interesting. There are people in America (many) who simply write "Caucasian" or "White" on forms that ask what their ethnicity is. While this may be true, they don't realize that they may have African or Hispanic countries in their ancestral background and possibly some Native American as well. In some cases, a significant percent of their ancestry may have actually come from the Middle East. The same is true of African people in America; they may have more European ancestry in their back ground than they could have ever imagined. So, breaking into those mysteries via scientific DNA testing that is proven to be accurate (although different companies may report analytical findings differently), can be actually be fun, important and truly gratifying. Learn more about DNA testing via my compiled postings; each being numbered, following.
JML

Who are You?

1. How My First Testing through Ancestry.com Compared to My Known Family History

I did ethnicity testing through an ancestry.com. They have a U.S. testing lab that is purported to be near 100% accurate in determining ethnicities, using many 1,000s of DNA markers. The program is headed by PhD scientists specializing in genomics and bioinformatics.

People provide vials of saliva mixed with a stabilizing/preserving solution and they send the collected sample to the DNA testing lab. The testing analysis is for maternal (mother's side) and paternal (father's side) ethnicity via "Autosomal DNA". Formerly, testing labs offered only the dad's Y-chromosome or the mother's Mitochondrial DNA testing which only gave maternal or paternal analysis but labs have now developed newer, each sex inclusive methods for testing both, called autosomal DNA.

The tests results for me show I'm of 47% Great Britain ethnicity and 27% Ireland. PLEASE NOTE: I will be getting re-tested through a different DNA lab, toward the end of September, 2015 (the new 2nd test is reported later in this book). This first test left me a 4th (26%) of my ethnicity from elsewhere in the world, which they determined the origins of on my remaining test results called "Trace Regions" and that I will list following.

I was disappointed that my rumored Native American DNA was simply not there <u>on this particular test</u> (it did show up on my 2nd test however). I believe the test results I will obtain in late September 2015 **will show otherwise** and if so, I will explain why I believe the new test is showing a small percent of American Indian.

Regarding my surname (last name) history, I have an MD cousin - a historian who tediously and systematically traced my "Lowance" surname ancestry - with two other researchers helping him, to the family of "Saint Lawrence of Rome" ("Lawrence" being the Latin version of Lowrance). Our ancestors were of either Italy, France or Spain (likely a mix).The name "Lowrance", literally means "Man of Laurentum" - a city on the west coast of the Italian Peninsula southwest of Rome, "the original capital of the Latins" according to Wikipedia.

It turns out that other parts of my slightly more ancient DNA ethnicity, does indeed include Italy, France and Spain These are areas the ancestry DNA testing people circled analytically, on European maps - one of which showed Italy centrally, another showing Spain centrally and a third showing France centrally.

France, Spain and Italy are also encompassed/included on the analytical maps of 'Great Britain', 'Italy', 'The Iberian Peninsula' and 'Europe West'. With these inclusions, this raises Spain (mapped 3 times), France (Mapped 4 times) and Italy (mapped 5 times) as origins of my ancestry, to a higher level.

Who are You?

So, am I part Spaniard, French or Italian? ...Probably all three but most of my DNA is British and Irish (because of current end-immigration), with Scandinavian and Finn (Finland) being included. There is a small amount of Asian mixed in as well - 2%.

2. My Reaction to the Analytical Ethnicity Maps

The 2% Asian DNA, with the analytical map showing India centrally and a large area of West China was a surprise! My highest percent ethnicity is Great Britain as mentioned previously, which didn't surprise me. France and part of Italy are also within the circle of the Great Britain analytical map, while these same countries are highlighted-centrally on others, which I found interesting and exciting. This was true of other maps as well, some of which also included Spain (highlighted centrally on one of the ethnicity maps).

I called the ancestry.com DNA testing people and they said even lighter-highlighted, circled areas represent actual DNA ethnicity. So, even Great Britain adds more French and Italian into the mix, being within the circled areas of my DNA. These countries are on the other mapped DNA areas as well (e.g. 'Europe West' and 'Iberian Peninsula'), the latter two mentioned, also showing 'Spain' and 'Italy' again. I felt that the combination of these facts showed very interesting ancestry in my background.

Some people might say "but no one has that many ethnicities in their DNA", however, some of these ethnic groups likely go back 100s of years. My very small amount of Asian-Indian DNA (2%) could have easily been a British soldier ancestor who married an Asian woman. There was rapid expansion of British power through the greater part of the Indian subcontinent in the early 19th century.

The Italian, European-Spanish, French, Scandinavian and Finland also go further back in my ancestry than does my UK and Irish DNA. Regardless, I still have some of the DNA from those ancestors, who over time, immigrated to different countries, intermarrying with these different ethnic groups. They eventually settled in Britain and Ireland for a long period of time and then moved on into the United States. I am a Caucasian-American although some of my ancestors were somewhat darker-skinned people (e.g. Italian/Greek, Spaniard and Asian-Indian). I think many of us suppose that 23andme and ancestry.com simply have DNA strands they use that are universally recognized, as a stand-alone method for determining origins of people, as far back as 6 generations (or up to 500) years.

However, one view is that they each want to be their own self-contained ancestor tree seeking companies. So, I believe they calculate, I'm sure via very qualified scientists and technicians, people's individual ancestries based on their own collected samples.

I also believe this is why companies like GedMatch and others, can take the actual raw data and show you a truckload more information (I've done this through them). Still I had to express wonderment as to why more Ancestry composition is not going to be added in the "New Experience" announced by 23andme.com for Europeans (or is it; will more be added?), rather than for Asian people and African people only (I also have both of these ancestries at different points in my own DNA but many others as well).

In short the two companies of Ancestry.com and 23me are a help with 'family tree' building (it spreads the service by specializing on this aspect). And, not as much help to those who only want complete DNA breakdowns (Ancestry Composition and Countries of Origin), which consists of matches from people and countries who have specifically provided them samples. Millions of samples can definitely help a project to match, batch and give breakdowns for DNA. They simply have decided that it's not the direction for the two competing companies to go. In regard to 23andme, this may prove to change for the better sometime after Nov. 2015 (I'm very hopefull).

3. My Second DNA Testing through 23andme.com

I wrote the chapter above shortly after receiving my results from a previous DNA test I completed with Ancestry.com and now, I want to show results of my new DNA testing via 23andme.com...

They are a reputable DNA/genome matching lab-company.

Starting with a bit of my own known background first, I was always told I have American Indian ancestry and my newest/second DNA testing seems to confirm that I do have Native ancestry. It plus another DNA testing lab, both also show Asian Indian in my ancestry, as well as Native American. However, reputable genome science reporting sites state that 1/3 of American Indian DNA comes originally from Asia. I look fully white/Caucasian (so-called) but I tan a bit more easily than other Caucasian people I know and my wife has said to me for the 32 years of our marriage (as of this year-2015), that I have somewhat of an olive colored skin (With light brown eyes and light brown hair, that used to be blonde in youth?). This doesn't indicate a sure thing of any type but the DNA testing does come far closer to a sure thing regarding Asian, especially when there are ethnicities found that match between 2 different DNA tests I took, by two different labs.

On my first testing, I had a solid 5% Italian but the newer test shows some of my percent of DNA as "largely southern European" and some as "largely northern European" (generalized - not categorized). I believe this may be where these same ethnicities are found. Iberian, France and Spain are listed as DNA finds on both tests, as well as a large percent to be UK and Ireland. In "The Countries of Origin" section on the new test, there's quite a big spread for my origins.

Italy is the 6th highest percent on this aspect, by the new test - among a couple dozen countries my ancestors have/do live in. This, in spite of Italy and other of these countries not being listed on the main results page of the more recent test by 23andme.com. I was actually amazed because quite a few Hispanic countries are among these as well, including Spain, Brazil, Mexico, Chile, etc.... In spite of what I state in some of my information above, there were a great deal of closely-matching ethnicities between the 2 DNA tests.

So my gene pools are highly spread-out but with the final landing places for many of them now being in the part of North America many of us call home - The United States of America. Many also settled for a long while in England and Ireland (my highest percent of modern, non-American DNA), with many I'm sure, still being in these countries. Does this mean I am not 100% white/Caucasian due to the non-white ethnicities?

I am mostly white period, but yes **not 100%**. Whatever the case however, it doesn't matter to me, I'm just happy knowing more about who I am and I'm proud to be an American. What really is white anyway - skin color? Some of my DNA shows Middle East origin as well. I'm of the "J2" Haplotype; a major result my other DNA test didn't include. This major genome from my paternal side is found in about half of Jewish men and even more Lebanese are of this J2 type Haplogroup but other Middle East peoples are as well.

So, that's basically what I have learned about my DNA ancestry. One final point I would make, is to point out that my surname: "Lowrance", was literally named after "The City of Laurels" in Italy (a type of tree that grew throughout the city). By the time my ancestor "Saint <u>Lawrence</u> of Rome" came along (his family – not necessarily directly from him), this was the Latin change to the name that eventually became "<u>Lowrance</u>". I have some typed-out – hard copy ancestry studies in my study desk, showing this same fact regarding St. Lawrence of ancient Catholic clergy that were compiled in the early 1960s. These tests have added more strong confirmation to these things written down by my relatives who researched.

I know little about the European period of my family, other than the fact of our being in the ancestral linage of the family of St. Laurence of Rome. My grandparents and the great and "great, great" ones, were field farmers and they raised farm animals. One group of the "greats", were Welsh and lived/worked on whaling ships. Some genome scientists say that many Welsh immigrated directly to Wales from Rome during the religious persecutions.

4. **My Complaint about "Southern Europe" and a 23andme Surname Affiliate Site**

If you look at my "3% Southern European" on 23adme.com test results, it's highly obvious that this particular report should be higher.

If you look at the "chromosome view" of my Ancestry composition for Southern Europe, strand number 3 on my dad's side, is about 50% South Europe (SE). Strand number 6 about 45% SE. Finally, strand number 8 shows about 10% SE on my dad's side. This should absolutely reflect more than "3.0% SE". I also cannot imagine this much SE not being able to be categorized (i.e. Italy, Greece, Balkan Peninsula or Sardinia); these are most of the more common countries/states represented by SE. I also have Asian ancestry, that shows up on only one strand (maybe 10% of it) and yet they showed it to be 2% of my DNA/Genome on test results. How can 3 strands, one being 50% color coded SE, another 45% and another 10%, all together be shown as "3.0% SE"? I mean no offense by this whatsoever but it is extremely obvious even to a layperson like me that the SE percent needs to be adjusted (common sense and not simply because I want it). I made a screen shot but there's no prompt on this PM form for sending it to the DNA testing company. I hope they can take a second look at it for their selves. Also as an added note let me say that the very name of our branch of Lowrance's, literally did have its beginnings in Italy, so "0.0% Italy" shown on the AC, would indicate that this surname history never happened, when it obviously did. Their affiliate site among many others, shows that it did: https://www.surnamedb.com/Surname/Lowrance . Additionally, I have family tree history books stating that my part of the Lowrance family, is the one that started in Italy as "Laurence". The next chapter discusses more about this.

Who are You?

5. Lowrance History – Some Recent and Some Ancient

My dad served in the military, toward the end of the Korean War. My oldest brother served during the Vietnam conflict. My great grandfather was a Sheriff Deputy, while he also farmed. My dad, being raised a farm boy was the 1st generation to earn college degrees when he was still a young dad (he is now age-79 as is my mother). He worked for a period of time at technical engineering companies such as General Dynamics and Texas Instruments.

He later settled into less technical work as a salesman for various companies including those that sold Specialty Foods and Health and Beauty Aids. Sales are what I held as my occupation as well, working for one of sales company for 18 years. I am also an author and using a search of my name online will take you to sites that show my book, eBook and audiobook titles. Some of my books I have written are about my successful invention marketing and licensing years (I'm still receiving royalties from one invention - since '96).

Other than graduating high school, I earned an accredited diploma from a well-known theology college (33 credits upon completion in 1996). My daughter is currently in the process of receiving her master's degree from this same college – to be completed in December if 2015.

She first received her bachelor's degree at a contemporary college. I was a youth minister for 20 years and was also in general ministries (teaching classes and congregations). My second-to-oldest brother is a church pastor and this has been his calling/occupation for close to 35 years (as of this year of 2015). My family is proud of his ongoing service to The Lord. In year 2012 I began receiving U.S. Social Security benefits for health disabilities. I write about my illnesses in many of my books. As a Christian, I believe in divine healing but that it is performed within God's timing and wisdom.

I recently found in my dad's library (mentioned previously and now in my own library), a book of history for my family surname (last name) "Lowrance". I was of course amazed at the information it contained, In fact it was this and one other book my dad had possession of, that inspired me to get DNA testing done. As previously-stated, I completed 2 different DNA/Genome testing lab tests, via saliva sampling (Ancestry.com and 23andme.com). These strongly confirm what is written in the book that was researched regarding the Lowrance family, their origins and the history of name changes made from "Laurentius" (Italian, French), "Lawrence" (Latin) and the end and present surname "Lowrance". Our present surname sounds ethnically less like these other past ethnicities and more like UK/English, New England and American, where we made our final immigrations to.

Regardless, when you look up the surname online, ethnicity sites state that Lowrance is derived from the Latin Lawrence which in turn is derived from Laurentius, meaning "City of the Laurels" and "from Laurentum" (a city near Rome). I'll now simply type out the research conducted in the 1960s by Earl and Rubye Littrell, below. I do not know how I'm related to this couple (if at all) who researched Lowrance past history (including surname origins) in major libraries throughout the USA, but I am grateful to them for having done so. I believe they have since passed-on but a cousin of mine wrote "PART TWO" and the remainder of the book in the 1960s – plus a genealogy book of his own about families connected to the Lowrances (by Peter B. Berendson).

My greatest interest is in the history of my Lowrance surname but my cousin named Peter Berendson in PART TWO, writes about families that became related-to and engrafted-into the Lowrance family (related surnames). With this said I will now type-out the first page of "PART ONE", following below (exactly as it appears in the 50 year old document – as of this year-2015).

The family tree book-page compiled in the 1960s. - - -

LOWRANCE

The Name

Lowrance drives from the Latin. The origin is ancient and honorable. A city of Rome was named Laurentum. Perhaps a century after Christ, there lived a Christian Martyr destined to be one of the most famous and celebrated Saints of the Roman Church. He took his name from his city and was known as Saint Laurentius. (Or Lorenzo). In England he is known as Saint Lawrence.

Saint Laurentius was called "The Deacon" and was known and revered as the friend of the poor. For almost three centuries after the death of our Lord, Christians were persecuted by the Roman Government at the express command of the Emperor, whichever one was on the throne, because the Emperor demanded to be worshipped as God. The entire empire was searched for Bibles, any Christian writings, and for anyone that would not bow down and kiss the royal ensign. Some Christians were fed to the lions in the Coliseum for sport by the ruling class. Emperor Nero, in his great persecution, covered Christians with bitumen, elevated them high on poles in his garden and burned them for light during his orgies.

Saint Laurentius is said to have been roasted on a gridiron in the year 258 during the persecution by Emperor Valerian. A shower of Meteorites appeared at this time and are known as "the tears of Saint Laurentius".

The heroic courage in the face of fierce persecution by these Martyrs won millions of converts to the Christian faith and Christianity emerged in a mighty triumph over the paganism of Greece and Rome.

The spread of the name

Christianity triumphed over Caesar. Emperor Constantine accepted Christianity. Rome ruled over all Europe: England, France, Germany, and the Baltic States; in fact, the known world. Evangelists were sent throughout the land, carrying the name and fame of Saint Laurentius with them. The name was adopted and adapted. A province of France was named Lorraine. The name in France was changed principally to Laurens or Laurent. The name in Germany became principally Lorentz or Lorents. In England the name became principally Lawrence or Lawrance. Under the influence of these changes and with the passage of centuries, other variants became common: Lowery, Laurie, Lawrie, Lawson, Lorance, Lowrance and others. *(End of 1960s family tree book page.)*

Me (JML) Again: The page I have reproduced verbatim above, written before online search was available shows my Lowrance surname to be of Italian descent.

A Roman Deacon – Saint Laurentius, whose family passed the name down, fled to France during Roman persecution, where they changed the name to Laurent or Laurens, hoping to avoid persecution had they instead carried the Laurentius name. As Christian persecution spread into France, Laurent or Laurens cover-names (for lack of better term), as a protestant group, fled into Bavaria and lower Germany and called themselves "The Huguenots". 1,000s of Huguenots were massacred by the non-protestant French army. Many fled to Holland, where they afterward embarked on America.

FINAL POINTS: During these persecutions described above, there were reformed-name ancestral family of mine that stayed and mixed with other ethnic groups. Some of the Laurentius family in Italy for example must have crossed the border of France, to Spain. Some are said in genetic studies to have traveled directly to Whales with some of the neutral Romans. Part of my UK is Welsh because other old genealogy books have shown this regarding my family; they lived and worked on whaling ships.

To repeat, both my Ancestry.com and 23andme.com show me to have some Asian as well (near or in India). Strangely 23andme shows "0.2%" Asian, while Ancestry.com shows "2%" (no decimal – ten times higher). Also Ancestry.com shows me to have 5% Italian, while 23and me.com shows 0.0% (??? but a true percent may be reflected in their "Broadly Northern or **Broadly Southern European**" categories).

Ancestry.com also shows me at 3% Iberian (basically Greco-Italian), while 23andme.com shows me at "0.1%" Iberian (30 X difference between the two, wow!?).

I am not knocking 23andme.com because they show what likely most white Americans do on DNA tests – that we are mostly UK and Irish (Our most recent heritage). According to the family genealogy books written in the '60s about my family, these are the final destinations of most European immigrants. For my family, this has only been the case for 2 to 3 hundred years. This leaves 100s and 1000s more years of ancestry-genealogy to be filled-in. I'm very grateful to 23andme.com for showing my Haplogroup genome "J2", which includes pre-Arabised Mesopotamians and Levantine peoples, Mediterranean/Aegean peoples, Greco-Anatolians, Caucasians, South and Central Asians (haplotypes is something Ancestry.com doesn't provide).

NOTE: Caucasians in this case means "the Caucasia people who lived between the border of Europe and Asia, and South and Central Asians." – According to Wikipedia. I will add a final point in saying that 23andme.com showed me to have Native American DNA, while Ancestry.com did not.

6. The Expected Changes coming soon to 23andme.com

Fellow ancestry seekers, be sure to read this post, you are sure to find it important and interesting (the small bit of my surname history I'm adding has a purpose). My disappointment as well as my area of praise for 23andme will be within this post.

The **COA** aspect (**Countries of Ancestry**), which will be partially or totally removed soon, has struck me a bit strange (but it may be replaced with a better version!). BTW: I downloaded mine just a day or two after results were in for me months ago. I also printed it to paper. This is no offense toward 23andme but for example, my Southern Europe shows 3.0%...and yet it also shows 0.0% Italian and all other Southern Europe countries also show 0.0%, with exception of Iberian Peninsula showing 0.1%. So, they can detect via my strands, an area I have 0.1% in but not the other 2.9%. How then did they find 0.2% Asian in my DNA!? Maybe it has to do with their number of in-house genome collections for matching purposes (?). Their coming "**New Experience**" (revamping available DNA information on their site), may prove to be an exciting and delightful change. If you're reading this after November, 2015 – the New Experience should be in place.

I'm sorry to add that on 23andme.com, I do not like the prompt on the ancestry composition page that gives the 3 options of "speculative", "standard" and "conservative".

As you change from "speculative" which already appears on the page, to these other two options, your ancestral composition becomes further and further diminished or even erased. On the "conservative" setting, I'm 4% "British and Irish", 62.5% "Broadly Northwestern Europe", 30.8% "Broadly Europe" (2.8% unassigned). Wow that's a huge variance from the "speculative" view.

What I am grateful for as stated earlier, is the Haplogroups 23andme performed analysis for on my test (I'm "J2" on my paternal side and H5 on my maternal). This is something Ancestry.com doesn't as yet perform for clients. Lastly there are a number of sites online confirming the origin and development of my surname as stated above, such as this one: https://www.surnamedb.com/Surname/Lowrance (a 23andme affiliate site).

I think each of our browser-types and versions of word-business programs, cause each of us different effects as far as "Microsoft Works" type spreadsheets go. Mine kept coming out as unreadable spreadsheets, so I have been going to the different pages and using my snipping tool, I go to the different 23andme pages and get the different things I need saved in files and printed on paper. So far this has worked well for me. If you don't have a "snipping tool", there are sites out there that allow you to download one and it's very easy to use.

I suggest using the main companies like Microsoft and avoid the middleman sites as they tend to install unwanted toolbars along with the snipping tool, even if you opt out of them (they act like a virus). Some PCs let you simply snip whole pages and you can resize if needed, using the "Page" prompt at top of your PC (usually left side labeled "Page"). Others may suggest even simpler methods for you.

23andme.com has gained more FDA liberties – if I am correct in what I have read about them but they are almost surely trying to avoid problems with them at the same time. I do look forward to seeing what they add to our experience on the site via the new updated site (after Nov. 2015). Hopefully it's not simply "the cap on allowed connections" (only 1,000 DNA relatives listed for us at any given time) and the inability for people to hide their DNA profiles (they must be visible to the 23andme public). Will there also be detailed DNA profiles with the changeover? This is still hard for me personally to determine via their announcements to us (Sorry to say so 23andme! – its likely lack of perception on my part). But, we shall see and I'm feeling optimistic!

7. Discussing Health Issues on the 23andme Forum

(NOTE: 23andme was at one time permitted to release DNA analyses that revealed risks for illnesses. They may again gain permission by the FDA to do so, which is another "wait and see" possibility.)

I have an autoimmune thyroid disease rare to men and more common in women. I also have diabetes with neuropathies, non-alcoholic fatty liver and non-smoker COPD (I'm only moderately overweight and had asthma as a child). I have also been found to have a nodule in my upper right lung but one more CT scan in Dec. this year of 2015, will hopefully rule-out the possibility of cancer. I will have had 2 years' worth of CTs in 6 month intervals – with no change in the nodule, which is good news. I also have severe arthritis and a common heart murmur called Mitral Valve Prolapse (mine has mild blood leakage as well) and I'm only age-52. There may be no DNA strands that show anything for these conditions or if there are, I may learn nothing from them if health DNA is allowed to be released. It's possible it may only be for people of child-bearing years, with a tested partner. I'll be interested in details of what this new aspect reveals.

Someone who is a member of 23andme.com, mentioned that some of us were getting **Countries of Ancestry** and **Ancestry Composition** confused with each other. I admit I am one of them however, "it's six of one and half a dozen of another" (I know, a very cliché saying). I used all features from day-one and the three of course I like best are CoA, AC and the Haplogroup. I believe it's possible that the AC will continue and not the CoA but I'm hoping some updated genetics, info-wise will appear when the new experience with 23andme.com is rolled out. On the other hand they may continue but, with improvements for each.

In regard to a 23andme Member who asked about thyroid problems, I offered a summation of my own thyroid disease story, which follows:

My thyroid disease started with Hyperthyroidism and transitioned to Hypothyroidism (hyper/high - hypo/low). Your TSH goes the opposite of your thyroid hormone levels. High TSH = hypothyroidism, Low TSH = hyperthyroidism. I take 2.5grain (150mg) of MD-prescribed Armour Thyroid® USP, going on 11 years now. Fixing my hypo, also raised my flagged-low Vitamin D levels and helped to normalize my then borderline diabetic glucose levels as well..

Thyroid antibodies should also be blood tested for (A Thyroid Antibodies Panel). This would be the "anti-TPO" (Anti-thyroid peroxidase) and the "anti-TG" (Anti-thyroglobulin). These, one or both, when found above normal range, indicates autoimmune thyroid disease - the type that cause destruction to the gland and sometimes at the same time, causing inflammation that enlarges the gland (Goiter). The anti-TPO is the one that most often elevates. Not in all cases however.

When I was diagnosed with thyroid autoimmunity (Hashimoto's thyroiditis), my anti-TPO was "120" in a normal range of '0 – 35' and my antiTG was a whopping "537" in a range of '0 – 40'. I've known people with thyroid antibodies in the 1,000s, which can risk of thyroid cancer and a condition called "Thyroid Storm".

The latter-mentioned being a severe condition that can be fatal or severely damaging to the body if not treated promptly. You can have thyroid disease with thyroid hormone & TSH in normal levels but elevated antibodies and symptoms of a sick thyroid (i.e. fatigue, musculoskeletal pain, stomach problems, etc...)

Lastly you should have the "TSI" antibodies tested (Thyroid Stimulating Immunoglobulin), which is found in high levels, above normal with "Graves' Disease" (autoimmune hyperthyroidism).

8. Do Names on your DNA Relatives Discovery List Change on Occasion (Question directed at me)

Yes I have seen the names change on the CSV spreadsheet. I like scrolling through the "DNA Relatives" section of my pages. I have "18" 2nd cousins and "886" 3rd and 4th cousins. I can easily screenshot/snip these but I'm not sure it would do much good at this point. I have so far only done screenshots (along with the ancestry composition), of the DNA color-graph with the associated countries. I have also enjoyed looking at the "Map View" of the DNA Relatives section. I will be screen shooting - making snips of those, which I have neglected to this point. This will be just in case they are affected by the coming changes at 23andme.com.

I went into my CoA page after all my results were in some months ago and snipped the CoA, as several jpg.picture files.

The "5" setting they give option for via the "Advanced Tools (sliding bar)", shows 40 or so potential ancestral countries for me with some as low as "0.1%". I took these snipped segments but couldn't snip them all at once. When I originally tried to do so, the colored-squares-graph was in the way. I then pasted them onto a word.doc page, including one with the graph on it. It all fit on ONE PAGE once I sized them. I printed a hard copy of them (totally readable), plus will keep them in my picture files.

All four of my grandparents were born in Oklahoma. I was born in Texas but raised in Oklahoma, where I still reside. My great-grandparents are a different story, having been born in Whales England. When I look at the birth and death certificates of my ancestors and at their boarding on ships to America (only 20 or so of each), all are either French or Italian names. I found that I'm 2% Eastern Indian via Ancestry.com results and this was confirmed with a smaller percent by 23andme, with their adding of Native American to the puzzle, I figured when much of Asia was under British control at one time, this explained it. Some Italians of Roman citizenship, traveled with Romans directly to Whales. As they probably spent time in England and becoming British citizens, there was inevitably intermarrying with Asian women. So, with us non-pedigrees, the trace amounts of huge numbers of countries could have easily happened. Still I like seeing the significant percent of different ancestor origins.

Who are You?

I like what Ancestry.com did. They updated their site and the countries they circle color-coded, branch into other circles that are more lightly colored, then to areas that are circled but have no color within that part of the circle.

For example my 47% Great Britain says this:

GREAT BRITAIN
Primarily located in: England, Scotland, Wales
Also found in: Ireland, France, Germany, Denmark, Belgium, Netherlands, Switzerland, Austria, Italy

My 5% Italy says this:

ITALY
Primarily located in: Italy, Greece
Also found in: France, Switzerland, Portugal, Spain, Serbia, Hungary, Bulgaria, Austria, Croatia, Bosnia, Romania, Turkey, Slovenia, Algeria, Tunisia, Montenegro, Albania, Macedonia, Kosovo

In regard to 23and me.com, my "J2" haplogroup would indicate middle east ancestry, so here again the widespread mystery deepens. I may also have Jewish or Lebanese ancestry or "admixture ancestry". So we who have lots of different DNA findings, can - instead of calling ourselves human-mutts can simply say we're highly non-pedigree☺.

If I am to believe the CoA (Countries of Ancestry), I have a "0.9" for Italy on it.

This, with my Advanced Control set to "5cM" (five centiMorgans – a measurement that catches distant relatives) and yet the AC (Ancestry Composition) shows "0.0" Italian at 23andme.com. Ancestry.com shows 5% Italy and 3% Iberian Peninsula. At 23and me, the CoA has a manual sliding level that changes the number of CoAs and the AC has specifically 3 levels of speculation/belief you can use to base your ancestry on. You may have in your reality-ancestry background, the weakest, strongest or middle confidence level shown. You actually chose this yourself which causes a wide, seemingly somewhat unreliable spectrum. I did in fact attempt to download the CoA when my results were first all-in (completed in several steps) months ago, via the CSV spreadsheet but it ran words together and my prompts to change it or view individual sections, didn't help. I did eventually find a way to do this however.

What I did instead was to use my snipping tool, as I related earlier and I copied as many countries with percentages applied to me as I could, in sections until I got all of them at the "5cM" level. It shows me to have 41 countries, with about half dozen being Hispanic, the Bahamas, etc... (I never would have guessed). So, of the two, I prefer the AC (Ancestry Composition) over CoA but of course I would like to have both. I do hope when they do the switch that changes needed to the AC, if any, will be done (I'm optimistic).

9. When I started with GedMatch.com

I took the advice of a member at one of the DNA platform's forums. As they directed me, after telling me about **a 3rd site** how I could do cross DNA matching and admixture DNA ancestry on. I went to the site "GedMatch.com" and was shown how to upload my "Raw DNA" and both Haplogroups from 23andme, and ancestry.com into their DNA research site (zipped files). I saw where I could donate a small fee to them but this is optional – not required. I am thrilled to have been referred to them. After learning about huge numbers of admixture DNA information they can provide (countries of DNA mixed with other ancestry countries), I paid an optional fee gratefully because this would allow me to learn even more about my ancestry! I checked-out the sites research credentials and the scientists involved literally numbering in the 100s (if not 1,000s), were impressive.

Their research is extensive, including using deceased peoples DNA (decades of studying ground bone, hair follicles, marrow, etc.), to provide ancient DNA and a far-reaching ancestry picture. Yes, they go back many 1,000s of years in some cases (archaic) but also show DNA ancestry that's far more recent. The number of DNA specimens researched by the different DNA science groups involved; both from living and dead specimens, is impressive (each of the scientists have their own websites as well). They place all of the information into one online computer source. The studies were in some cases, many years in the making by the geneticist-scientists involved.

Users of the platform, are directly involved in the processing of the information, by picking individual countries you wish to know whether or not contains ancestral DNA matched to you. Or, you can bring up charts that are simply calculated from your Raw Data sources, which is combined with their research studies involving the very same DNA strands/segments, in addition to your haplogroups. These combinations provide amazing results. One available prompt allows you to simply "triangulate" your DNA tests with your haplogroups and this alone provides surprising results. They also show you where your two or more DNA tests by different companies, varies in their information provided to you. There were "14" differences between my ancestry.com and 23andme.com tests but the mismatch calculation lines were very small in each case. So, the raw data by the two companies, matched relatively closely.

My belief is that their provided reports are skewed by a number of things and they can be reported slightly incorrectly. The raw data however, is still there. It's possible they fear being accused of sensationalizing results, so they limit countries or combine them as one result, etc..... Example: Why does 23andmecom have the "Speculative", "Standard" and "Conservative" options I mentioned earlier, all having very different displayed results on your Ancestry Composition page?

This 3rd company I have resorted-to, also provides me 1,000s (no exaggeration) of DNA matched segments, making them close, moderate or distant relatives (each provides their email address by choice); depending on the amount of matched DNA strand-segments. One fellow poster to this thread pointed out to me that you can't match relatives like the family of "Saint Lawrence of Rome" (my supposed ancestor described earlier) and that is absolutely true.

I will have to take the 2 genealogy books in my family's possession, with a grain of salt so-to-speak. What you can only do is recognize name similarities and the ancestral trek made by the supposed ancestor when looking at written historical inforation. All of the countries traversed/settled by the relatives of St. Lawrence of Rome, were on my results pages, which allow you to place the names of countries and the DNA test you choose (raw data in my case from either 23andme or ancestry.com) to have searched. I usually looked at results from each test separately and results were always very close to matching (To Repeat: My haplogroups play a major factor). My results did show France - South West, Italy, Iberian, Greece, Southern Europe and Spain, many times on both of the test's raw data analyses, combined with my J2 and H5a1 haplogroups.

The surprises were Eskimo, Brazilian, Norwegian and African (Nigeria, Ethiopia, Kenya, and Nile-Sahara). None of it matters to me, I think it's all cool.

One might say the 3rd company I'm using is not legit but this could be said of the other 2 companies as well but my opinion is that all 3 companies are legit but the 3rd one I have used (GedMatch), has far more info available to users of its platform. I sincerely believe that. The fact is that without the first 2 tests, I would have had no search capabilities at this 3rd website.

One thing that bothers me slightly is 23andme combining countries together. For example(s), my "British-Irish" category and my "French-German" (this may change with their new site layout after Nov. 2015). Ancestry.com separates these out. There ARE relatively modern boundaries for countries today and I know beyond a doubt that 23andme.com can separate them out as well. They were able to extract "0.1" of Iberian Peninsula on mine and yet anything else in Southern Europe is called "2.9% Southern Europe". Only a very few countries are involved - so, why no detail? When I use GedMatch.com, I get **detail galore** - not only on direct percentages but also on admixture and secondary population's analyses.

Absolutely every aspect of DNA research/discussion has people who are very pro or very con on all related subjects. What I'll personally say about **haplogroups** (I'm not saying this applies to you who are reading), which are very diverse among large populations of people, is that they zone-in on the very place where your ancestors branched-out. Having no starting place of larger population groupings, makes all other DNA compositions less-meaningful.

I have a list of about 900 people who I share large segments of my DNA with (countries of ancestry), who are alive today and many are in the same haplogroup areas shown in my DNA. Many ancestors from only a few generations back, stay in haplogroup areas of peoples (it's true from my own paternal and maternal sides) for 100s or 1,000s of years and they didn't immigrate to the USA or other countries until the past 200, 100 years or even less (others – not at all). For me personally, this is one reason that makes the haplogroups interesting, as offered by 23andme.com and that can be ran through GedMatch.com as well, for deeper analysis.

According to GedMatch.com, I have a percent of DNA from all the countries in the Iberian Peninsula. My opinion....23andme.com and Ancestry.com, whom I both appreciate (regardless of that sounding contradictive), have put some vagueness into the Ancestry Composition because they offer other programs including "experts" you can purchase the services of, to bring out a better picture of your ancestry for you. At the same time 23andme announced their "New Experience" coming, their price for kits went up from $99.00 to $199.00 correlating with the announcement. The new kit and analysis **may very well be worth the price**. If the 23andme new experience is overblown when announced, we will know this soon (I suspect however, that we will all be very happy with it). I snipped their letters and they promise the new experience to people who have already purchased kits and are registered on their site.

If the experience really is an improvement (I believe it will be), I personally will openly express my praise for them. I will give them a very favorable review.

10. My Expressed Disagreement with 23andme.com over a Key Area Regarding Countries

Following is the email I wrote them:

This is James Lowrance Test Kit Barcode: XXXXX. If you look at my "3% Southern European", it's highly obvious that this particular report should be higher. If you look at the "chromosome view" of my Ancestry composition for Southern Europe, strand number 3 on my dad's side (top one), is about 50% South Europe (SE). Strand number 6 about 45% SE. Finally, strand number 8 shows about 10% SE on my dad's side. This should absolutely reflect more than "3.0% SE". I also cannot imagine this much SE not being able to be categorized (i.e. Italy, Greece, Balkan Peninsula or Sardinia); these are among the few countries/states represented by SE. on Ancestry.com, I have Asian, that shows up on only one strand (maybe 10% of it) and yet they showed it to be 2% of my total ancestry composition. How can 3 strands, one being 50% color coded SE, another 45% and another 10%, all together be shown as "3.0% SE" on 23andme.com? I mean no offense by this whatsoever but it is extremely obvious even to a layperson like me that the SE percent needs adjusted.

Who are You?

I made a screen shot but there's no prompt on this Personal Message form for sending it to you. I hope you can look at it for yourself (thank you!). Also as an added note let me say that the very name of our branch of Lowrance's, literally did have its beginnings in Italy, so "0.0% Italy" shown on the AC, would indicate that this surname history never actually happened. Your affiliate site among many others, shows that it did: https://www.surnamedb.com/Surname/Lowrance . Additionally, I have family tree history books stating that my part of the Lowrance family, is the one that started in Italy as "Laurence". Thanks so much for any attention you can have done on this. NOTE: Once I ran my 23andme test kit trough GedMatch.com, the Italian numbers were revealed to be much higher, due to the fact that the site goes further back in generations as far as Ancestry Compositions that members of the site are given.

11. My Favorable Review for 23andme.com

I found a DNA Matches prompt on 23andme, that gives the names of any living people currently found related to you (the site user), unless they list their selves as "anonymous", which was under the "Match Name" heading. The other two headings were "Maternal Grandfather Birth Country" and "Maternal Grandmother Birth Country". That's only two generations away (the list is a little over 1,000 lines long).

I'll list some countries where these grandparents were born - usually both, which was the case for the vast majority.

Mexico, Italy, France, Finland, Chile, Bosnia-Herzegovnia, Croatia, Austria, Iran, Russia, Lithuania, Brazil, Netherlands, Spain, Belgium, Denmark, Japan, Guatemala, Portugal, Belize, Ukrain, Greece, N. Korea, Equatorial-Guinea, Singapore, Serbia, Romania, Czech Republic, Barbados, Philippines, Equator, Turkey, Cuba, Bahamas, Columbia, Sweden, India, Slovakia, Malta, Bulgaria, Venezuela, Hungary. (About 41 countries).

With these people having DNA from the countries listed only 2 generations back (maternal side), why is it that only France/French shows up on my 23andme "DNA Ancestry Composition", out of all of them (knowing they bunched together within much of Europe)? One might answer..."It's because they aren't related to you in regard to those countries you list, that both of their grandparents are from but rather from their British DNA for example, if they have some."

So, I'm to believe this in each case? Especially with the fact that the "Countries of Origins" tool stated **"These are countries you have DNA in common with"** and the program listed many of the same countries that are listed on my hard copy of this "DNA Matches" program. People's names and email addresses were on this "DNA Matches" list but I left emails off when I printed-off the list.

Who are You?

This was so that my sheets can't be used improperly by anyone else who might find them.

Keep in mind that these countries I listed above, don't show up just one time each. Some of them showed up dozens of times (Italy 46, Spain 11, Brazil 14, etc...), others 8, 6, etc..., I don't remember a single stand-alone country listed.

When I started with a different test originally (Ancestry.com), I simply wanted more ancestral information than I received from them via my saliva test sample after their analyzation of it. I had already been hearing about 23andme for years, namely about the other DNA testing company. So, it was actually a coin-toss type decision to try the other company first. When I decided to also use 23andme for ancestral DNA purposes later on, I at first thought I received practically the same thing from another simple saliva test. However, as I browsed the website, I began finding more and more very helpful information. This included the page showing my haplogroups on both the paternal and maternal sides of my family. These revealed a great more about my ancestry, than the DNA Ancestral Composition did alone. The haplogroups show even more about your ancestry and there are secondary DNA project companies online that allow you to download your raw DNA data to them and they will go deeper with the data. All of them I have seen suggest using your 23andme raw data for deeper analysis and that it is the best available.

They do this by extending-out the generations of your ancestors (further back in time) and confirming that your DNA data is correct.

Home DNA testing is still in its infancy. With this being true, the GedMatch project is head and shoulders above the rest because they not only go much further back in ancestry (23andme.com only does so with "Neanderthal") but they also show much more detail on DNA compositions in general. That includes more recent generations like Ancestry.com and 23andme.com show; up to 500 years or up to 6 generations but also much further back. I can load the same raw data from either of these two family tree sites (or both together), at GedMatch and my ancestry composition and countries of ancestry will be far more detailed and separated. I never expected as much from the family tree DNA sites but a bit more separation on the compositions would have been nice. Maybe that's part of what they're doing now, rather than doing this for Africa and Asia only - hoping (BTW: I do have both African and Asian admixture ancestry, shown many times at GedMatch).

Now, here's where my appreciation for 23andme comes in! Without the raw data DNA they got from my sample and them having determined my haplogroups, I would not have this info to run through GedMatch! This company even checks their raw data for errors and neither of my raw data files from Ancestry.com or 23andme.com had any. So, was the kit from 23andme.com worth the 99.00 I paid? Yes, in fact it would have been worth the now-price of 199.00. I know I sound contradictory but it's the best way I know how to express these things without sounding like I'm casting nothing but aspersions on either of the family tree companies.

Not everyone needs the additional raw data analysis conducted by other genome sites, but I did. They do not require an additional DNA sample from you - only raw data you already have available from your testing sites. I did the further analysis (not possible without first having sample analysis through 23andme.com). I was simply more interested in as many ancestry details as I could get. Other 23andme.com members (probably most) are interested in building family trees with the 6 or so generations of ancestral DNA their analysis reports give them and certainly that is also a worthy pursuit.

In short, I have become far happier with 23andme, since finding the many prompts on their website that provide more information for a **country-origins guy like me**, which I formerly overlooked when I first began browsing the site! Now I look forward at the end of November 2015 of benefiting even more from 23andme.com, once they go live with their improved DNA/Family Tree website!

(END)